Helps and Hints for Science in the Beginning

by

Dr. Jay L. Wile

Helps and Hints for Science in the Beginning

Published by
Berean Builders
Muncie, IN
www.bereanbuilders.com

Manufactured in the United States of America
Fifth Printing 2019

ISBN: 978-0-9890424-1-3

Cover Photos © Jason Keith Heydorn, Willyam Bradberry, Galyna Andrushko, Inga Ivanova, mozzyb, solarseven, and Artens via Shutterstock, Inc.

Cover design by Kim Williams.

All Scripture verses are taken from the New American Standard translation.

Helps and Hints for Science in the Beginning

Introduction

Thank you for using *Science in the Beginning*, a multigrade elementary science course. I have designed the course so that elementary students of all ages (K-6) can work through it together. The details of how that is accomplished are contained in the introduction to the main book. In this companion book, you will find the answers to the student review exercises, helpful hints about some of the lessons, tests, and answers to the tests. I pray that these contents will make the course easier for you to cover.

As I mention in the introduction to *Science in the Beginning*, while I do provide tests in this book, I personally don't think they are very important for the elementary years. However, I realize that there are many parents who do. In addition, I understand that some students need practice taking tests in science before they are thrust into the more academically challenging science courses found in junior high and high school. As a result, I ask that you use your own discretion when it comes to administering the tests. Do so if you think it is the best thing for your children. If you decide to use the tests, please note that you are free to photocopy them straight from the book. There is a note to that effect on the relevant pages.

If you are just looking for an evaluation tool, I would strongly recommend that you emphasize the student notebook. Each lesson has a notebooking exercise for all but the youngest students. These students should have a single notebook that is devoted solely to this course, and they should use the notebooking exercises as a guide for what belongs there. Of course, the notebook should be the student's own account of what he or she learned in the course, so the student is free to add things to the notebook. The notebooking exercises simply indicate the minimum content that should be found there.

How will you evaluate your student's notebook? Obviously, the student's completion of the notebooking exercises should contain scientifically correct information, which is why you will find answers to all of the exercises in this companion book. If the student ends up putting incorrect information in the notebook, have him correct it. That will offer a chance for the student to learn from his mistakes.

However, the notebook should be more than just a repository for information related to the course. You should also use it as a tool to help the student master his strengths and improve on his weaknesses. For example, does your student struggle with writing? The writing assignments in most of the notebooking exercises are fairly short, so you can view them not only as science work, but also as English and grammar work, perhaps going so far as to grade him on how well he writes. If writing is a real struggle at first, consider having your student orally answer the exercises, with you writing down what he is saying. As time goes on, you can transition from him doing none of the writing to him doing more and more of the writing. If your student is good at writing, then view the notebook as a chance to encourage him to be creative with his words,

Is your student good at drawing? There are several drawing assignments in the notebooking exercises. Make part of the student's grade based on how good the drawings are. Alternatively, if your student is not very good at drawing, there are options in most of the assignments to cut pictures out of magazines or print them off the internet instead. Allow your student to do that *most* of the time, but force him to draw in a few assignments, just so he has a chance to work on that weakness. Of course, if drawing is a weakness, any grade you give based on his drawings should take that into account.

For some of the lessons, there are hints given, usually regarding the experiment. It is always best to look at a lesson's entry in this book before starting the lesson with your students so that you are aware of any hints that might make the lesson a bit easier. Also, if the student is asked a question during the reading, such as to identify something in a picture, the answer will be found in this book's entry on that lesson.

As discussed in the introduction to *Science in the Beginning*, there is a way for you to ask questions about the material. Feel free to use that service as often as you need. My goal is to make science more enjoyable for both you and your students!

Table of Contents

Lesson Review Helps and Hints

If your children are having trouble completing a lesson review, check here for help. You should do the checking, not your students.

Lesson 1

1. Light reflects when it bounces off something, changing direction.
2. In order to see something, light must reflect off it and hit your eyes. If there is no light, that can't happen.

Older students: See #1 above.

Oldest students: The drawing should show an arrow coming down from the light and hitting the object on the table. Where that arrow hits the object, another arrow should start, and it should end up hitting the person's eye. This allows the person's eye to detect the light and send information to the person's brain so the brain can form the image. If the light were turned off, the person would see nothing.

Lesson 2

1. They came from the (mostly) white light made by the candle flame.
2. Red, orange, yellow, green, blue, indigo, violet

Older students: See the pictures on page 5. Mr. White Light's name helps you remember the colors of the rainbow because each letter stands for a color, and the letters are in the same order that the colors appear in the rainbow.

Oldest students: You would not see the rainbow. Only white light has all the colors of the rainbow.

Lesson 3

Note for the experiment: The black paper will reflect some light, because it cannot absorb all the light from the flashlight. As a result, you will see some of your helper's paper when you shine the flashlight on black paper.

1. It reflects red light.
2. They are absorbed by the object.

Older students: In the drawing, the only arrow that should reflect off the rose and hit the person's eye is the red arrow. The seven arrows represent white light, which has all the colors of the rainbow. The red arrow represents the only color of light that is reflected – red.

Oldest students: See the "older student" explanation. The other colors are absorbed.

Lesson 4

Note for the experiment: If the two pieces of plastic aren't at noticeably different temperatures when your students touch them, just let them sit in the sun longer.

1. Light is a form of energy.
2. When an object absorbs light, the object gets warmer because the light's radiant energy is changed to thermal energy.

Older students: The lesson discussed radiant energy, chemical energy, mechanical energy, and thermal energy. The picture for radiant energy should use light, chemical energy should use food, wood, or something else that contains chemicals, mechanical energy should have something in motion, and thermal energy should have something hot.

Oldest students: A black shirt converts the radiant energy of the sunlight into thermal energy, which warms you up. The white shirt reflects most of the radiant energy, so it is not converted into thermal energy.

<u>Lesson 5</u>

1. A white object reflects <u>most</u> of the light that hits it.
2. The white shirt still absorbs *some* light, and that light gets changed into *some* thermal energy.

Older students: When the magnifying glass was held over the paper, it concentrated a lot of sunlight into a small space. That means there was a lot of radiant energy there. Even though the white paper reflected *most* of that radiant energy, some of it got absorbed and got changed into thermal energy, which was enough to catch the paper on fire. If the paper had been black, it would have caught fire sooner, since it would have absorbed more light.

Oldest students: The one on the left is a mirror. It reflects almost all the light that hits it, and all the light reflects in the same direction. The middle one is black paper, since almost all of the light is absorbed. The one on the right is a white piece of paper, since almost all the light is reflected, but it is scattered all over the place.

<u>Lesson 6</u>

Note for the experiment: If you can't see the light bulb light up, see if the balloon has enough charge by touching the top of the balloon to the wall. If the balloon sticks to the wall, it has enough charge. If not, you need to use cleaner hair or a wool blanket.

1. Energy cannot be <u>created</u> or <u>destroyed</u>. It can only be <u>converted</u> from one form to another.
2. In order for you to have fun, your body needs to make a lot of mechanical energy. The only way you can do that is if you have chemical energy, which you get from food. If you don't have a lot of chemical energy in your body, it won't be able to make a lot of mechanical energy.

Older students: Energy cannot be created or destroyed. It can only be converted from one form to another. In the experiment, the mechanical energy in the balloon's motion was converted into radiant energy. Since you could only put a little bit of energy into the motion of the balloon, it got converted into just a little bit of radiant energy, which is why the bulb was dim.

Oldest students: Moving the balloon quickly would give you a brighter glow, because there is more mechanical energy in a quickly-moving balloon than in a slowly-moving one.

<u>Lesson 7</u>

1. A battery stores chemical energy.
2. There are only so many chemicals in the battery, so there is only so much chemical energy. When the chemicals get used up, there is nothing more that can be converted to electrical energy.

Older students: The batteries convert chemical energy to electrical energy, and the motor in the toy car then converts electrical energy into mechanical energy.

Oldest students: The car will not run as fast with two good batteries instead of three, because there isn't as much chemical energy to convert to electrical energy. With one battery, it will run even more slowly.

<u>Lesson 8</u>

Note for the experiment: The key is that the digital camera needs to be of low quality. High-quality digital cameras filter out infrared light.

1. We call it visible light.
2. You cannot see infrared light or ultraviolet light. The students need to name only one.

Older students: The drawing should show an arrow coming from the remote, hitting the paper, and reflecting off the paper and to the television, which is behind the remote. If someone stood in front of the remote, it wouldn't work, because that person would be blocking the light.

Oldest students: The proper order is infrared light (lowest energy), visible light (medium energy), and ultraviolet light (high energy).

Lesson 9

1. It focuses the light on the retina.
2. They detect light.

Older students: The drawing should look very similar to what is on page 25. The rods and cones are on the retina, and the blind spot is that brown area where the optic nerve comes out on the retina.
Oldest students: The green circle will disappear when the book gets close to your face. When the light coming from the green circle hits your blind spot, your brain has to "fill in" the missing information with what it sees around the green circle, which is just white paper. So the green dot is replaced with white paper.

Lesson 10

1. The light can reflect off the object, be absorbed by the object, or pass through the object.
2. It means that light can pass through it.

Older students: The first drawing should show light hitting the fork, reflecting off the fork, traveling through the water and then the air, and then hitting a person's eye. In the second drawing, the arrows should show light hitting the fork, reflecting off the fork, passing through the water, reflecting off the surface of the water, passing back through the water, passing through the bowl, and hitting a person's eye.
Oldest students: The students should have an explanation like what is written above.

Lesson 11

Note for the experiment: If you know that you can yell at each other and be heard through the window, there is no reason to work out signals.

1. You will see your reflection.
2. When a large amount of light is doing one thing and a small amount of light is doing another, we tend to see what the large amount of light is doing.

Older students: If your little brother has his lights on, the small amount of light coming from the flashlight outside will be overwhelmed by the reflection of light inside. As a result, either your little brother won't see anything, or it will not be nearly as ghostly looking.
Oldest students: For a one-way mirror to work, the side on which the suspect is being questioned needs to be brightly lit. The side from which the other officers are observing needs to be dimly lit. That way, the amount of light being reflected on the suspect's side is large, and all you see on that side is the reflected light. As a result, the window looks like a mirror. On the other officers' side, very little light is reflected, because there is little light to begin with. That means the majority of light comes from the room where the suspect is, so the officers see the suspect very clearly.

Lesson 12

1. It reflected off the air.
2. It is called a fiber optic cable.

Older students: The assignment explains what the drawing should look like.
Oldest students: The doctor can put a tiny camera on a flexible tube. She can push it in the patient's mouth, down the esophagus, and into the stomach. The camera will need light, though, which is where fiber optic cable comes in. A fiber optic cable can be pushed right along with the camera, and light will shine wherever it goes. This is often done in diagnostic medicine.

Lesson 13

1. Morse code converts every letter in the English language into dots and dashes.
2. You need the chart to convert your written message into flashes of light, but your helper needs it to convert those flashes of light back into a written message.

Older students: This was a simple code, but it still converted information (the answer to a question) into flashes of light, and the light was what sent the information to your helper.
Oldest students: You could push a fiber optic cable through the crack. Then, you could convert the password into Morse code. You could then use flashes of light to send that Morse code through the fiber optic cable. You wouldn't necessarily need a flashlight, but that would be great. You could just turn the lights in the room on and off, and most likely, enough light would travel through the fiber optic cable to be seen.

Lesson 14

Note for the experiment: The amount of milk is very important. Without enough, it will be hard to see the beam of light in the water. With too much, the water will be too cloudy to see the beam.
1. Light travels more quickly in air.
2. In order for you to see the underwater part of the pencil, light had to travel from water to air at an angle. This caused the light to bend, which made the underwater part of the pencil look like it was in a slightly different place.
Older students: The first drawing should look like the picture on page 40. The second should show just one beam, since there is no bending. Refraction is the bending of light as it passes through a transparent object. It depends on the object and the angle at which the light hits the object.
Oldest students: The more tilted the flashlight is relative to the pan, the more the light in the water will bend.

Lesson 15

1. A magnifying glass is made from a curved piece of glass.
2. It changes shape to change what light gets focused on the retina.
Older students: The puddle of water was curved, so light bent depending on the angle at which the light hit the puddle. That concentrated light, making the puddle a magnifier.
Oldest students: A curved piece of glass acts as a magnifier for the same reason a curved puddle of water does. A flat piece of glass will not act as a magnifier, however, because the light coming from the object below it hits the flat piece of glass straight on. As a result, there is no bending and no concentrating of light.

Lesson 16

1. A cloud is made of water in its liquid (or sometimes solid) phase that has formed on small particles floating in the air.
2. Ice is water in its solid phase. The water that you drink is in its liquid phase, and when water evaporates, it turns into its gas phase.
Older students: The three phases of matter are solid, liquid, and gas. Ice is water in its solid phase. The water that you drink is in its liquid phase, and when water evaporates, it turns into its gas phase. A cloud is formed by the evaporation of water from the rivers, lakes, streams, and oceans of the earth. The water vapor rises and then condenses onto tiny particles in the air. The water can eventually freeze, if it's cold enough. A cloud, then, is liquid (or sometimes solid) water that has formed on particles floating in the air.
Oldest students: Many Christians think the expanse is air because air separates the water in the rivers, lakes, streams, and oceans of earth from the water in the clouds. In addition, the Bible says God called the expanse "heaven," and in Hebrew, that can mean "sky," which is made of air.

Lesson 17

1. The main difference is that solid water takes up more room than liquid water.
2. In order to get something to melt, you must add thermal energy.
Older students: The square representing solid water should be *larger* than the square representing liquid water, because water expands when it freezes. The square representing solid wax should be *smaller* than the square representing liquid wax, because wax contracts when it freezes.

Oldest students: The copper would take up more room when it melted, because almost everything in creation except water contracts when it freezes. Thus, liquid copper would take up more room than solid copper.

Lesson 18

1. Volume measures how much room an object takes up.
2. In order to sink, an object must weigh more than an equal volume of water.

Older students: In the drawing, the larger square should be floating, and the smaller square should be at the bottom. Since the student was told to assume both squares had the same weight, then the one that sinks must take up less volume, since something that sinks must weigh more than the water that it must shove out of the way.

Oldest students: It would have to weigh more than 400,000,000 pounds. Since the ship floats, it must weigh less than an equal volume of water. If the ship weighs 400,000,000 pounds, that means an equal volume of water must weigh more than that.

Lesson 19

1. Ice floats in water.
2. Most lakes would freeze solid during a long, cold winter.

Older students: The story should have the fish living in the lake through many winters, just avoiding the ice that sinks to the bottom. However, during the long winter, the ice on the bottom will keep piling up until there is nowhere left for the fish to swim. If the child wants a happy ending, someone could notice the fish's plight and bring it to an aquarium inside.

Oldest students: If you could collect an equal volume of water, all you would have to do is weigh the object and weigh the equal amount of water. If the object weighs more, it will sink. If it weighs less, it will float.

Lesson 20

1. Evaporation occurred in the saucepan, while condensation occurred on the upside-down lid.
2. The upside-down lid represented a cloud, because that's where condensation occurred.

Older students: The drawing should illustrate the description, which should include the fact that water evaporates from a body of water, travels up until it is cold enough to condense. The water condenses on small specks of dust, and as time goes on, this forms a cloud. The cloud moves, collecting water from other places, until it gets so heavy the water must fall back to the earth as rain.

Oldest students: If the air is cold all the way to the ground, the water will fall as snow, because it stays solid the entire time. If the air is warm, the water will fall as rain, because it is not cold enough to turn to its solid phase.

Lesson 21

Note for the experiment: The pan will be heated with nothing in it for a while. Some pans don't react well to that, so make sure you have a pan that won't be harmed.

1. Water is cohesive because it is attracted to itself.
2. Water is adhesive because it is attracted to other things.

Older students: The explanation should include the fact that there was a layer of water vapor under the water. That kept the water from touching the pan and thus kept the pan from heating the water enough to boil it. The water moved around the pan because some of the thermal energy that the pan was giving to the water vapor was converted into mechanical energy. Only one drop was left in the end because water is cohesive, so the little drops eventually came together to make one big drop.

Oldest students: Adhesion is the tendency of water to be attracted to other things, while cohesion is the tendency of it to be attracted to itself. If water is strongly attracted to what it is on, the adhesion will win, and the water will not bead up. If water is not strongly attracted to what it is on, the cohesion will win, and the water will bead up.

Lesson 22

Note for the experiment: The battery won't be good to use in electronics after the experiment.
1. You would have a molecule of water.
2. The electrical energy destroyed water molecules, turning them into hydrogen and oxygen.

Older students: The drawing should show few (or no) bubbles in the glass with no Epsom salt, more bubbles in the glass with some Epsom salt, and the most bubbles in the glass with more Epsom salt. The explanation should include the fact that the electrical energy from the battery broke the water molecules apart, forming hydrogen gas and oxygen gas, which is what made the bubbles.

Oldest students: The model shows two white balls (two hydrogen atoms) and one red ball (one oxygen atom), so the drawing should have two hydrogen atoms linked to an oxygen atom. Water is called H_2O because it consists of two hydrogen atoms and one oxygen atom.

Lesson 23

1. Sugar is the solute and water is the solvent.
2. The sugar would be left.

Older students: The story can be as creative as the student wants. However, the main "plot" should be that the ions start out close to one another when they are in the saltshaker as a solid. However, as the salt is dissolved, the water pulls them away from one another. The student needs to indicate somewhere in the story that the salt is the solute, the water is the solvent, and the two things form a solution.

Oldest students: Baking soda will dissolve in water (but use only a little). Any powdered drink mix will dissolve in water. Other solids would include sugar, tea (best to use a tea bag), and artificial sweetener. In each case, the solid is the solute, and water is the solvent. The solution is what results (sugar water, baking soda and water, etc.).

Lesson 24

1. Carbon dioxide is dissolved in soda pop.
2. Gases dissolve better in cold water, so it was harder for the carbon dioxide to leave the solution.

Older students: The explanation should include the idea that the pits in the Mentos attract carbon dioxide so it can form bubbles. It should also mention that something in the Mentos (gum arabic, but the student needn't name it) makes it harder for the water to hold onto the carbon dioxide. It should mention that those bubbles form rapidly and rise rapidly to escape. This pushes Diet Coke up into the air, making a fountain. A warmer bottle would make an even larger fountain.

Oldest students: Air must be pumped into the aquarium to dissolve more oxygen into the water. The fish use up the oxygen that is dissolved, and unless more oxygen dissolves to take its place, the fish will run out of oxygen and will suffocate.

Lesson 25

1. Air pushed down on the cork.
2. Wind is made up of air.

Older students: The explanation should include the fact that air was trapped in the glass. Because of that, it pushed down on the water that was in the glass as well as the cork, causing the cork to end up at the bottom of the sink. This shows that air exists because of the effect it has on other things.

Oldest students: The cork would not get pushed down to the bottom of the sink if the glass had a hole in it. Air could escape through the hole, so it wouldn't have to push down on the cork.

Lesson 26

1. It expands.
2. The speed of its molecules increases.

Older students: The explanation should include the fact that air expands when it is heated, so it expanded to inflate the balloon. If the student wants to get more detailed, the explanation can include the idea that the molecules in the air start moving faster as the air in the bottle heats up, hitting the balloon harder and causing it to inflate.
Oldest students: The pictures just need to illustrate the effect. I personally would draw two pictures: One of a bunch of balls (representing molecules) that have lines behind them, indicating motion. These "motion lines" indicate that the balls are moving many different directions. The caption would read "colder air." The other would have the same balls, but the motion lines would be longer, indicating that they are moving faster. The caption would read "warmer air." However, as long as the explanation provided makes sense, the drawings can depict anything that illustrates the idea.

Lesson 27

1. The Bible tells us that air has weight.
2. The balloon full of cold air weighs more.

Older students: The explanation should include the fact that air has weight, but that because air expands when it is hot, hot air weighs less than cooler air. When the heat is turned up, then, the balloon gets bigger and eventually weighs less than an equal volume of the air surrounding it, allowing it to float. When the heat is turned down, the air cools, the balloon shrinks, and eventually the balloon weighs more than an equal volume of the air around it, causing it to sink.
Oldest students: The key concept here is that the *total weight* of the balloon *and* the air inside it must be less than an equal volume of air. If the balloon material weighs a lot, it will need to have a larger volume in order for it and the hot air inside to weigh less than an equal volume of air around it.

Lesson 28

Note for the experiment: The card needs to be strong enough to hold against the weight of the water, but it also need to absorb a bit of water. A non-laminated greeting card works as well as an index card.

1. Scientists call it air pressure.
2. The air presses on the book from all sides, which cancels out its effect.

Older students: The explanation should include the idea that the weight of the water and air (represented by the arrow pointing down at the card) is less than the strength with which air pressure is pushing up on the card (represented by the arrow pointing up at the card). As a result, the card cannot be pushed down from the glass.
Oldest students: The bottle will look like it has been crumpled. When the bottle was open at the top of the mountain, it had air in it, but the air was at a lower pressure. Since the lid was put on the bottle, air could not enter (or leave) the bottle. As the person drove down the mountain, the air pressure increased. The air pressure inside the bottle could not increase, however, so the bottle would crumple due to being pressed on the outside with more pressure than what is on the inside.

Lesson 29

1. Wind blows from areas of <u>high</u> pressure to areas of <u>low</u> pressure.
2. Differences in temperature cause those differences in pressure.

Older students: The drawing should have arrows pointing up from the plate to indicate the hot air that is rising. The ashes should be "lifted up" by those arrows. There should also be arrows pointing from the sides to the bottom of the plate, indicating air that was rushing in to replace the air that rose.
Oldest students: In general, colder temperatures produce higher pressures, and warmer temperatures produce lower pressures. Thus, considering only temperature and not altitude, the highest pressures will be found at the poles, where it is coldest. The lowest pressures will be found along the equator, where it is warmest.

Lesson 30

Note for the experiment: The key is to get an airtight seal between the straw and the bottle.
1. The water had to expand when it got warm, and the straw was the only place it could do that.
2. Scientists call them units.

Older students: The explanation should include the idea that when the water and air inside the bottle got warmer, they had to expand. The only way they could do that was to move water up the straw. When the water got cooler, they contracted, which made the water move down the straw. If there was a leak, air could be pushed out through the leak, allowing the water and air to expand inside the bottle without pushing water up the straw.

Oldest students: The other unit that is probably most common is the Kelvin. Water freezes at 273 Kelvin and boils at 373 Kelvin. The Rankine temperature scale is another one. Water freezes at 491 Rankine and boils at 671 Rankine. The Réaumur scale is another. Water freezes at 0 Réaumur and boils at 80 Réaumur.

Lesson 31

1. When a leaf starts to rot away so that it looks like dirt, we say that the leaf is starting to decompose.
2. It is the decomposed dead things found in soil.

Older students: Something decomposes when its molecules break down, and it becomes something quite different from what it was originally. Decomposed dead things are what make up humus, and humus is an important part of soil. The more humus, the richer the soil.

Oldest students: Humus obviously has something that is good for plants, since they grow better when there is more of it. In fact, this is the way God recycles things in His creation. Plants need certain things to survive. They get some of those things from the plants that are now dead. When the living plants die, new plants will be able to use their decomposed remains to grow. (If the student didn't get this answer, don't worry. It was designed to make the student think.)

Lesson 32

1. Soil is also made of small bits of rock.
2. Pores are tiny spaces in the soil.

Older students: When the water absorbed by a rock freezes, it expands, pushing against the inside of the rock. This happens over and over again, and eventually all that pushing breaks the rock. Soil is made of humus and bits of rock that have different sizes.

Oldest students: If you look at the soil that is underwater, you should see that the pores are larger in the soil near the bottom of the jar. As the text told you, those are the bigger bits of rock. So smaller bits of rock produce very small pores, while bigger rocks produce bigger pores. A mixture of different sizes of rock gives you pores that are between large and very small, which makes for a fertile soil.

Lesson 33

1. When sediments harden, they form sedimentary rock. When melted rock freezes, it forms igneous rock.
2. Metamorphic rock

Older students: The drawing should look like the one on page 101 and the explanation should be similar to the explanation next to the drawing.

Oldest students: Marble and slate are both metamorphic rocks. Sandstone and shale are sedimentary rocks. Granite and obsidian are igneous rocks.

Lesson 34

1. You find saltwater in the ocean.
2. A gallon of ocean water would be heavier.

Older students: The drawing should have two layers, with the freshwater layer on top. There can be some blurring in between the layers, but the layers should be noticeable. The explanation should indicate that freshwater weighs less than an equal volume of saltwater, which is why it floats on saltwater.

Oldest students: The can of Coke would eventually float. Remember, the can of Coke wasn't a lot heavier than the water. If you start dissolving sugar in the water, you start adding the weight of the sugar to the water. As a result, the sugar water starts getting heavier. Eventually, the can of Coke will weigh less than an equal volume of sugar water, and the can of Coke will float.

Lesson 35

1. When you dissolve something in water, it lowers the freezing temperature. So the freshwater will freeze at the higher temperature.
2. It is made of frozen freshwater, because it came from a glacier.

Older students: The story can be as creative as the student wants, but it should involve the snowflake being squished by the weight of snow and slowly flowing down the mountain, eventually reaching the ocean. The higher temperature causes melting, and the part of the glacier the snowflake is in eventually falls off the glacier and into the ocean. The formation of frozen seawater requires a lower temperature, because saltwater freezes at a lower temperature than freshwater, and a glacier is made of freshwater.

Oldest students: The definition should say the equivalent of, "When a solute is dissolved in a solvent, the freezing temperature of the solution is lower than that of the solvent." To determine where the ice came from, taste it. If it has a salty taste, it is a chunk of frozen seawater. If it doesn't taste salty, it is from an iceberg. Remember, the chunk was rinsed thoroughly, so any ocean water that was clinging to it should be gone.

Lesson 36

1. There will be *some* liquid the next morning, because the water never completely freezes. That's why salt can melt ice – there is some liquid there to begin with.
2. Salt melts ice better than sugar.

Older students: The explanation needs to have the idea that there is some liquid water there when the salt is put on the ice cube, because there is an equilibrium between water melting and refreezing. The addition of the ice slows the speed at which the water refreezes, so more ice melts than water refreezes. This makes more water for more salt to dissolve into. This doesn't work at very low temperatures, because if it gets cold enough, even saltwater freezes, and a new equilibrium can be established.

Oldest students: The bucket is an example of equilibrium because the bucket is losing water through the hole but gaining water from the faucet. When the level of water doesn't change, you know it is losing an equal amount through the hole as it is gaining from the faucet. That's equilibrium.

Lesson 37

1. It is called a hypothesis.
2. It means that the experiment did not turn out as you expected.

Older students: The answer needs to be that hot freshwater melts the ice more quickly, and the explanation should be because the cold freshwater floats on the hot saltwater, making it harder for the hot saltwater to warm up the ice cube.

Oldest students: To make sure your hypothesis is reliable, do more experiments. A single experiment might just give you the wrong answer because of a mistake you made. If you do several experiments and they all confirm your hypothesis, there is a better chance that your hypothesis is correct.

Lesson 38

1. The cotyledons were the biggest part of the seed.
2. It is called the embryo.

Older students: The drawing should look something like the picture on the bottom of page 114. The cotyledons are food for the embryo, and the embryo becomes the plant.

Oldest students: Some plants have only one cotyledon. It turns out that plants can be separated based on how many cotyledons their seeds have. Monocot plants have one cotyledon in their seeds, and dicot plants have two. The bean plant, then, is a dicot plant.

Lesson 39

1. The radicle is the first thing that emerges from a seed.
2. The root system anchors the plant and absorbs water for the plant.

Older students: The first thing that emerges from the seed should be labeled the radicle. The longest root should be labeled as the primary root, as it has been growing the longest. The other parts of the root system are the branches. In the steps of germination, "1" should deal with the testa starting to come off. "2" should discuss the radicle emerging from the seed to form the primary root. As it grows, it forms branches, producing a root system.

Oldest students: The hypothesis is up to the student. In a few lessons, the student will learn whether or not the hypothesis is correct.

Lesson 40

1. It pushes its way to the surface so the plant can get above the soil.
2. It develops into the stem.

Older students: The third step is the formation of the hypocotyl, which pushes up and out of the soil, bringing the rest of the plant with it. The hypocotyl eventually becomes the stem, which hold the leaves in the air and allows water to travel from the roots to the rest of the plant.

Oldest students: The hypothesis is up to the student. In the next lesson, the student will learn whether or not the hypothesis is correct.

Lesson 41

1. It becomes the plant's first true leaves.
2. They are called seed leaves because they do the same thing during germination that the leaves do later on – they feed the plant.

Older and Oldest students: The fourth step is the hypocotyl breaking above the surface and straightening to become the stem. The fifth step is the plumule emerging from the cotyledons and forming the first true leaves. During this step, we see the epicotyl, which is between the cotyledons and the true leaves. The sixth step is when the cotyledons run out of food. They whither and eventually fall off the plant.

Lesson 42

1. It is called photosynthesis.
2. It needs light, water, and carbon dioxide. It really needs chlorophyll as well, but it makes the chlorophyll itself, so the student doesn't have to mention that.

Older students: The description needs to include that the plant uses light (which is captured by chlorophyll), water, and carbon dioxide to make glucose, which is the plant's food. Oxygen is also produced and released in the air. For the plant drawing, the roots absorb water (and whatever is dissolved in it), the stem passes that water to the leaves, and the leaves make the plant's food.

Oldest students: Plants don't absorb food in their roots. They make their own food with photosynthesis, so fertilizer is not food. Fertilizer is plant vitamins and minerals. It helps to make the soil richer in the vitamins and minerals plants need to be healthy.

Lesson 43

1. They turn it into starch.
2. It turns the iodine dark blue or even black.

Older students: The explanation is that iodine turns dark when it touches starch. Thus, anything that turns iodine dark has starch in it. The bread's starch and cracker's starch come from the grains from which they are made. The potato's starch comes from the potato plant, because that's where the plant stores its food. The paper's starch comes from the paper-making process. The green banana's starch comes from the banana tree. It is there so that it can be converted into sugars as the banana ripens.
Oldest students: If the presence or absence of starch is surprising in something the student tests, you could do some research to find out why it does or doesn't have starch in it.

Lesson 44

1. They call it phototropism.
2. They call it gravitropism.

Older students: The box experiment drawings should show the plant changing the way it grew so as to grow towards the hole. The explanation should be that plants grow so that their leaves point towards the light so they can make as much food as possible. The scientific word is phototropism. The other experiment drawing should show the plant growing sideways and then curving so its leaves grow upwards. This is called gravitropism, and plants do it so their leaves grow up towards the sun and their roots grow down into the earth.
Oldest students: There are several: chemotropism (turning in response to chemicals), heliotropism (turning in response to the sun's position in the sky), hydrotropism (turning in response to water), thermotropism (turning in response to temperature), and thigmotropism (turning in response to touch or contact)

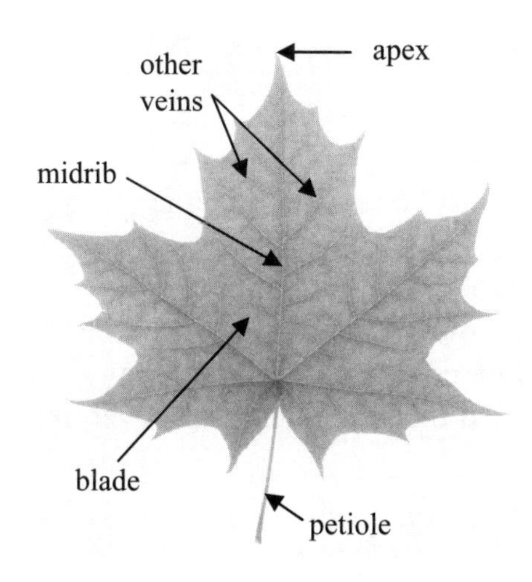

Lesson 45

1. It is darker because it has more chlorophyll.
2. The veins take water to all the different parts of the leaf and carry food from the leaf to the stem so it can get to the rest of the plant.

Older students: A labeled picture is show on the left. The top of the blade is darker because it has more chlorophyll. The bottom has more stomata so the leaf doesn't lose as much water when they are open.
Oldest students: The veins should roughly match what is shown on page 136.

Lesson 46

1. The height of the sun in the sky affects the length of a person's shadow.
2. Its shadow changes both length and position, like the toothpick's shadow did.

Older students: The drawing should have three lines coming from the dot. They should be at different angles, and the relative sizes should be the same as the markings on the paper used in the experiment. The length of the shadow changed because the height of the sun in the sky changed, and the position changed because the angle at which the light from the sun hit the toothpick changed.
Oldest students: The shadow will start out very long and will get shorter and shorter. However, by noon (the student doesn't have to say anything about noon – he can just say that at some time), it will be as short as it can be. After that, it will start getting longer again. This is because the sun gets higher

in the sky, casting shorter shadows, until noon. That's its highest point. After that, the sun starts sinking in the sky, casting longer shadows.

Lesson 47

1. A sundial uses a gnomon's shadow to tell time.
2. It reaches the highest point it will reach during the day at noon.

Older students: Because the position of the sun in the sky changes regularly over the course of the day, the position of an object's shadow changes regularly throughout the day. A sundial uses an object's shadow to tell how far the day has progressed, which is a measure of time. The sun reaches the highest point it will reach each day at noon.

Oldest students: The gnomon is the thing on a sundial that casts a shadow. It needs to be angled to account for the fact that the path the sun takes across the sky changes as the seasons progress.

Lesson 48

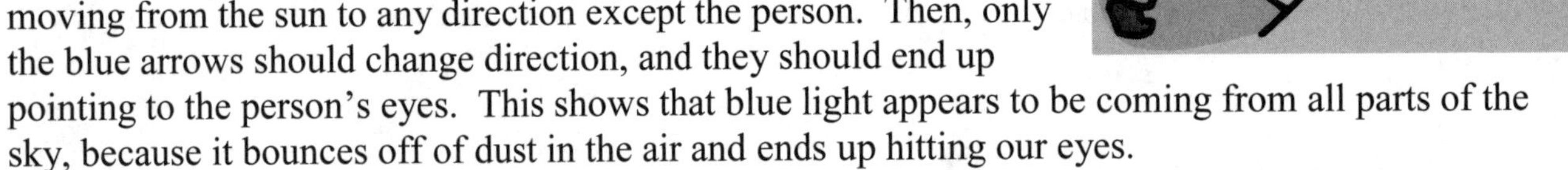

1. Blue, indigo, and violet light bounce off the dust in the air a lot.
2. The sky would be black, even during the day.

Older students: The drawing should look like the one on the right, showing that only the red, orange, and yellow light reaches the person's eyes. That's why sunsets have red, orange, and yellow in them.

Oldest students: The second drawing should look a lot like the drawing on the right, but this time, the arrows should start out moving from the sun to any direction except the person. Then, only the blue arrows should change direction, and they should end up pointing to the person's eyes. This shows that blue light appears to be coming from all parts of the sky, because it bounces off of dust in the air and ends up hitting our eyes.

Lesson 49

1. The earth's rotation causes the sun to move across the sky.
2. Like a smooth airplane ride, you just don't feel it because you and everything around you is traveling right along with the smooth rotation of the earth.

Older and Oldest students: The first picture should have a circle drawn around the earth, and the motion lines should show the sun moving along that circle. The second picture should just have the sun and earth, but the motion lines should show the earth spinning. Both could turn night into day because they both allow the sun to be shining on one side of the earth and then the other. However, the one with the earth spinning is the correct one.

Lesson 50

1. It takes a year, or just over 365 days.
2. The date for Easter is based on the lunar calendar, and we use a solar calendar.

Older students: The picture should show the earth spinning as it orbits. We mark the passage of the year with one full orbit around the sun. We mark the passage of a day with one complete rotation of the earth. The earth rotates just over 365 times in one orbit.

Oldest students: The other months have 30 days. Think about it. It takes over 365 days for the earth to orbit the sun, so our years are 365 days and 366 days. Thus, to make up for the fact that the event takes a little longer than 365 days, you add a year that is one day longer. To make up for the fact that the sun takes over 29 days to go through its phases, you add a month that is one day longer. If the student's answer is wrong, have him go back and write the correct answer along with the proper explanation.

Lesson 51

Younger Students: Any phrase in which each word starts with the letter representing a planet in the proper order would work. Some examples would be "**M**y **V**ery **E**legant **M**other **J**ust **S**ent **U**s **N**achos," "**M**y **V**ery **E**asy **M**ethod **J**ust **S**peeds **U**p **N**aming," "**M**y **V**ery **E**arly **M**orning **J**ust **S**tarted **U**nder **N**ightfall," or "**M**other **V**ery **E**asily **M**ade **J**elly **S**andwiches **U**p **N**ow."
Older students: The drawing should look something like the one on page 154.
Oldest students: Look at the size of each circle in the drawing on page 154. The circles for Mercury and Venus are much smaller than the circle for earth. That means they don't have to travel nearly as far as earth to get all the way around the sun, so they orbit the sun in much less than an earth year. Mercury, for example, takes only 88 days to orbit the sun. The planets farther from the sun than earth (Mars, Jupiter, Saturn, Uranus, and Neptune) have to travel a lot farther than earth to get around the sun, so they take longer than an earth year. Neptune, for example, takes 165 earth years to make one trip around the sun!

Lesson 52

1. Jupiter is the largest planet, and Mercury is the smallest one.
2. Stars make their own light; planets do not.

Older students: In order of size, the planets are Mercury, Mars, Venus, earth, Neptune, Uranus, Saturn, and Jupiter. Stars are larger than planets, and stars make their own light, while planets do not.
Oldest students: The inner planets are the ones inside the asteroid belt, and the outer planets are the ones outside the asteroid belt. The inner planets are smaller than the outer planets, and the inner planets are rocky, while the outer planets are big balls of gas.

Lesson 53

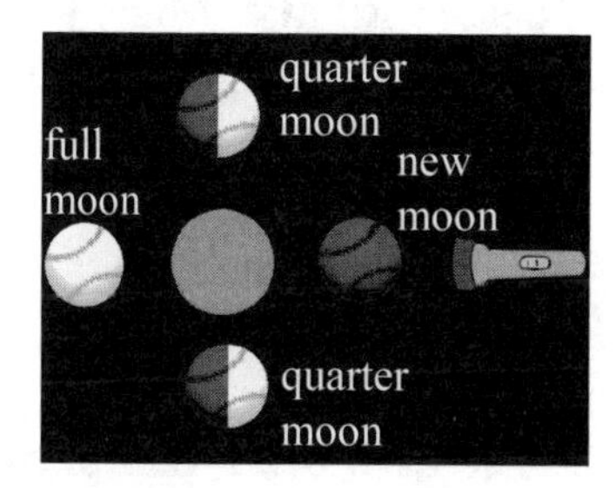

1. We call them the phases of the moon.
2. The moon orbits the earth.

Older students: The drawing ought to look something like the drawing on the right
Oldest students: The phases should look like the pictures of the moon on as shown in the figure on page 162.

Lesson 54

1. The star is actually bigger.
2. It is not closer to the earth – it just appears bigger because of an optical illusion.

Older students: The moon appears larger than the stars in the night sky because it is closer to us. However, it is larger when it is on the horizon because of an optical illusion that is created by your mind comparing it to other things you see near the horizon.
Oldest students: Put your finger in front of your face until it just covers up the pole. That way, your finger appears to be as tall as the pole. Now position your finger so the bottom is at the bottom of the tree. Mark where the top of your finger is with your other hand, and move your finger so that the bottom of your finger is where the top was. Continue to do this over and over until the top of your finger reaches or is above the top of the tree. Then multiply 6 by the number of times you had to do that. If it took you four times to get to the top of the tree, for example, the tree is about 24 feet high.

Lesson 55

1. The moon gets in between the sun and the earth, blocking the sun's light from part of the earth.
2. The earth gets in between the sun and the moon, blocking most of the sun's light from the moon.
Older students: The drawings should look something like what you see on the right. The gray and black circles in the top drawing illustrates that there are two regions to a solar eclipse – the region that sees a total solar eclipse (black), and the region that sees a partial solar eclipse (gray). A solar eclipse should be explained as the moon blocking the sun's light from hitting parts of the earth. The lunar eclipse should be explained as the earth blocking most of the sun's light from hitting the moon.
Oldest students: Because the earth has to be between the sun and moon for a lunar eclipse to occur, the moon is always where it is during its full phase – where the sun shines light on the entire side that faces the earth.

Lesson 56

Note: In the picture at the top of page 170, Sirius is the bright star on the left edge of the photo.
1. The farther away a star is from earth, the dimmer it will appear in the sky.
2. Apparent brightness is how bright the star appears in the night sky. Absolute brightness is how bright the star is if you are right next to the star.
Older students: The brightness with which a star appears in the night sky depends on both its absolute brightness and how far it is away from earth. Stars that are close to earth will appear much brighter, so a dim star close to earth could appear to be brighter than a bright star far from earth. The sun appears so large because it is so close to us. The other stars are farther away, so they appear smaller, even though many of them are larger.
Oldest students: Many stars that we see in the night sky are made up of more than one star. For example, while the dog star (Sirius) looks like a single star in the sky, it is actually made up of two stars that orbit each other very closely. Sirius A is the name of the larger star, and Sirius B is the name of the smaller one. Alpha Centauri is the same. It looks like one star, but it is actually made up of two stars that orbit each other – Alpha Centauri A and Alpha Centauri B.

Lesson 57

1. The light from the sun overwhelms the light from the stars. They are still there, however.
2. Light pollution is light that comes from human-made structures. It makes the stars in the night sky harder to see.
Older and oldest students: You don't see the stars during the day because the sun produces so much light that it overwhelms the light coming from the stars. The stars are still there, but they just cannot be seen. Light pollution is light that comes from human-made structures. It makes the stars in the night sky harder to see because some stars are so dim that even light from a city can cover them up, just as light from the sun covers all of them up.

Lesson 58

1. It absorbs ultraviolet light.
2. It is in the ozone layer, high in the air above even the tallest mountain.
Older students: Sunscreen absorbs ultraviolet light but allows visible light to pass through. This protects us from sunburns, because ultraviolet light harms skin and causes sunburns. If the ultraviolet light never hits the skin, a sunburn can't form. Ozone does the same thing. It absorbs ultraviolet light but allows visible light to pass through it. That protects life on earth. Most of earth's ozone is in the ozone layer, high in the air above even the tallest mountain.

Oldest students: The position of the ozone layer is important because ozone is poisonous to us. So earth needs a lot of ozone, but it needs to be where no one is breathing. If we could transport ozone, we should move it from the air we are breathing and put it in the ozone layer. That way, the air we are breathing would not be bad for us, and the ozone layer would block even more ultraviolet light.

Lesson 59

Note for the experiment: The sunspots will be like little specs of dirt inside the white circle that is the image of the sun. There may only be one or two. It is possible that the sun has no visible sunspots on the day you are looking, but that is unlikely.
1. The sun is a ball of gas.
2. Sunspots are areas on the surface of the sun that are cooler than other areas.

Older students: The explanation should include that sunspots are areas on the surface of the sun that are cooler than other areas. The number of sunspots was tracked as early as 800 years before the birth of Christ.

Oldest students: The relationship between sunspot number and temperature is counter-intuitive because sunspots are cool parts of the sun. You would therefore think that the more sunspots there are, the cooler the sun would be, but that is not the case. Since more sunspots mean a hotter sun, you can look at ancient records of the number of sunspots to get an idea of how hot the sun was back then.

Lesson 60

1. We can determine what is in it by studying the light that comes from it.
2. Emission spectroscopy is the study of light emitted by something so we can know what it is made of.

Older students: When chemicals are given thermal energy, they convert it into radiant energy. Each chemical gives off its own unique mixture of different types of light. If we use emission spectroscopy to study the light given off by the sun, we see the mixtures that are given off by hydrogen and helium. Thus, we can conclude that hydrogen and helium are in the sun.

Oldest students: A given chemical gives of its own unique mixture of light, which produces a specific color. The specific color given off by neon is a bright red/orange, as shown in the picture on page 182. This means that all neon signs used to be red/orange. (As a side note, nowadays we use other gases as well as special coatings on the tubes to produce a wide range of colors.)

Lesson 61

1. A Venn diagram helps us to compare two things, showing what they have in common and what their differences are.
2. An organism is a living thing.

Older students: The plants should have a lot of green in them and should be rooted to one spot. The animals should have legs, fins, or something else that allows them to move. They could also be eating.

Oldest students: There are lots of exceptions. For example, purple beech trees have red/purple leaves, not green ones. There are other plants that have no green during a good fraction of their lives, like trees that lose their leaves in the winter. The text told the student about Venus flytraps. There are other "insect-eating" plants as well.

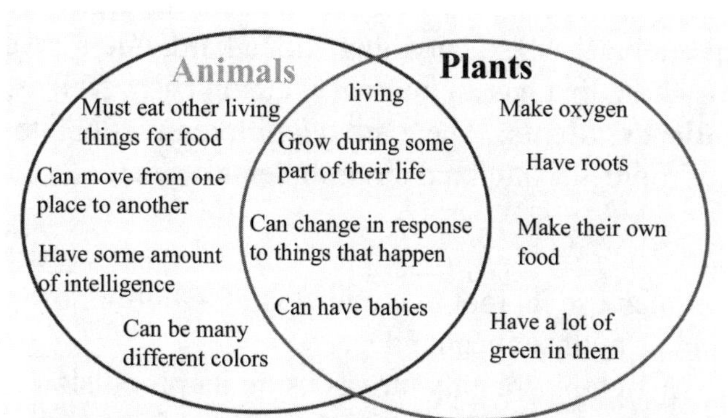

This is the completed Venn diagram. Note that when studying day 3, the students learned that plant embryos were baby plants. Also, plants respond to changes in light, such as the bean plant growing to the hole in the box in lesson 44 and the plant changing the direction it was growing when its cup was tilted in the same lesson.

Also, coral and sea anemones are animals, but they don't move easily. As the text says, some sea slugs can make their own food, at least part of the time.

Lesson 62

1. We call it osmosis.
2. Freshwater fish urinate a lot, because they have to get rid of the water osmosis is pulling into their bodies.

Older students: The first drawing should just have a fish underwater. The drawing representing a freshwater fish would have the fish expanding (or even blowing up!) because osmosis pulls water into its body. If it didn't get rid of the water, it would get bloated by it. The drawing representing a saltwater fish should show the fish shriveling up and dying, because osmosis is constantly pulling water out of its body. If it didn't continually drink water, it would shrivel up due to water loss.

Oldest students: Nothing would happen to the water in the fish's body. If the amount of solutes is exactly the same in the water and in the body, there wouldn't be a net pull of water into our out of the fish.

Lesson 63

1. Solutes like to travel to where there is a lot of solvent and very little solute.
2. A fish uses gills to breathe underwater.

Older students: A jellyfish is covered with a semipermeable membrane that allows oxygen to travel straight into its body because of diffusion. A fish has gills that have blood vessels with a semipermeable membrane. It pulls water over those gills so that diffusion can push oxygen into the blood vessels. Because the only way to get oxygen from the water is through diffusion, every animal that is always underwater must have a semipermeable membrane in contact with the water. Otherwise, there would be no way for diffusion to push oxygen into the animal.

Oldest students: Because solutes travel from where there is a lot of solute to where there is a little, if you just let the powder sit in the water, the powder would move from where there is a lot to where there is only a little. If you wait long enough, this will eventually even out the mixture so there is the same amount of powder dissolved everywhere. The problem is that this takes a *lot* of time!

Lesson 64

1. Vertebrates and invertebrates
2. Vertebrates have a backbone, and invertebrates do not.

Older students: Anything that lives in the water and can't support itself (jellyfish, sea slugs, etc.) is an invertebrate. Also, anything with a hard, outer covering (crabs, clams, etc.) is an invertebrate. Anything that has an internal skeleton (fish, whales, etc.) is a vertebrate.

Oldest students: There are *a lot* more invertebrates on earth than vertebrates. This is true for water environments and land environments as well.

Lesson 65

1. It can use its foot to move along the bottom of where it is living, or it can use its foot to dig into the sand or mud so it can hide.
2. An octopus mainly moves using jet propulsion.

Older students: The octopus and clam should talk about how they both can crawl along the bottom, but the octopus can also use jet propulsion to swim, which the clam can't do. The clam should point out that it can dig into the sand or mud and hide, which the octopus can't do.

Oldest students: A jellyfish opens itself up like opening an umbrella, which lets water in. It then closes itself like closing an umbrella, which pushes water out behind it, propelling it forward. Thus, it uses a form of jet propulsion.

Lesson 66

1. It uses them to push itself through the water, steer, and keep itself rightside up.
2. It uses its swim bladder to control its depth.

Older students: The drawing should look like the picture on page 200. A fish adds oxygen to its swim bladder to increase its volume, causing it to rise in the water. It removes oxygen from its swim bladder to reduce its volume and sink in the water.

Oldest students: Many fish use their dorsal fins for protection. They have spines that pierce the skin of anything that touches the dorsal fins. If you have ever caught a fish and tried to grab it so that your hand is touching its dorsal fin, you might have felt how painful that can be.

Lesson 67

1. It has little hooks on some of its barbules that grab onto smooth barbules and hold them together. They are like the hooks and loops on Velcro.
2. It has them for insulation.

Older students: The drawing should look like the picture on page 203. The barbs connect to each other because of smooth and hooked barbules. The hooks grab onto the smooth barbules, holding the barbs together. The shaft is hollow to keep the feather light.

Oldest students: Down feathers are used to stuff pillows, because they are very soft. They are soft because they don't have much of a shaft and don't have hooked barbules to hold the barbs together.

Lesson 68

1. It spreads oil on them. The oil comes from the bird's preen gland.
2. Waterfowl are birds that like to be around water. Ducks, geese, and cormorants are examples.

Older students: A bird waterproofs its feathers by spreading oil on them. Oil and water don't mix because oil molecules are not attracted to water molecules. This means the oil forms a protective barrier, keeping out the water. The bird gets this oil from its preen gland. Waterproof feathers are important because if a bird's feathers get heavy with water, it cannot fly. If a waterfowl's feathers got too heavy, it might sink!

Oldest students: The one that is floating lower in the water is the cormorant, because it allows its feathers to soak up some water. This makes it heavier, causing more of its body to sink.

Lesson 69

1. There is more air pressure below the wing, which is why the wing gets pushed up.
2. Lift is the upward push a wing gets because there is more air pressure below it than above it.

Older students: The drawing should look like the picture on the right. The main point is that there should be a lot more dots below the wing than above, indicating that there is more air pressure below the wing. The explanation should say that the greater air pressure below the wing pushes up harder than the air pressure above the wing is pushing down, causing the wing to lift.

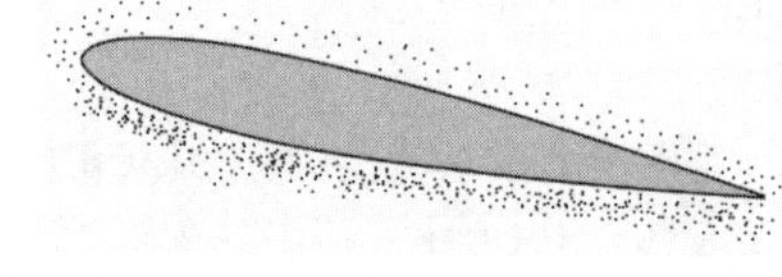

Oldest students: An airplane has to roll along the runway so that it can pick up speed. The faster it goes, the more difference between the air pressure above the wing and the air pressure below the wing. When the plane is moving fast enough so that the difference is great enough to provide the lift needed, the plane rises in the air.

Lesson 70

1. The umbrella trapped air underneath it, so there was more air below than above. That caused the air to push upwards, which popped the umbrella out.
2. It needs to reduce how much air it traps above its wings as they move up, because that creates a downward push.

Older students: The drawing should look a lot like the one for lesson 69, but the wing should look like a bird's wing. The bird brings its wing close to its body to reduce how much air it traps above its wings as they move up, because that creates a downward push.

Oldest students: Remember, if the wing is passing through air, lift is generated, because air travels across the wing, causing more air pressure below the wing than above. Well, if the bird is facing the wind, air will be traveling across its wings while they are stretched out, even if the bird is standing still. Thus, lift is being generated even when the wing is not moving. That will just add to the thrust generated by flapping! (This was a hard one. Don't worry if the student didn't get it.)

Lesson 71

1. It is shaped to reduce air resistance (or water drag) so that it can move through air (or water) quickly.
2. In order to fly, they must move through air. A streamlined body isn't as strongly affected by air resistance, which makes it much easier to fly.

Older students: Just like the Royal Tern, the airplane's body is also widest in the middle and narrow on each end. Just like a bird, an airplane needs a streamlined shape to reduce air resistance so it is easier to fly.

Oldest students: Even though a canoe doesn't have to overcome air resistance, it does have to overcome water drag, and a streamlined shape helps to reduce the water drag, making it easier for the canoe to travel through water.

Lesson 72

1. It reduces their weight, making it possible to fly.
2. No, they cannot.

Older students: The student should break open both bones and see which one is hollow. The hollow one will belong to the bird. The other will belong to the cat. Bird bones need to be hollow in order to reduce their weight so that they can fly.

Oldest students: The first bone is most likely from a bird. Because bird bones are mostly hollow, they are lighter than bones from other animals. Since the first one is bigger but lighter, it must be from a bird.

Lesson 73

1. All birds have bills. Not all birds have beaks. Only bills designed for striking and tearing are considered beaks.
2. Short, thick, and powerful bills are designed for cracking open seeds, so the bird eats seeds.

Older students: The bills need to look like the ones in the first five pictures in the lesson. The spoon-shaped bill tells us the bird scoops things out of the water for food. The short, thick bill tells you the bird eats seeds. The longer, thinner bill tells you the bird eats worms and insects underground. The sharp, curved beak tells you the bird eats animals that it catches, and the very long, thin bill tells you the bird eats nectar from flowers. They tell you these things because in each case, it is the best design for that kind of food. A bill is the general term for what leads to a bird's mouth. Only bills designed for striking and tearing are considered beaks.

Oldest students: A pelican scoops things out of the water to eat them. While the shape of the beak might not look exactly like a spoon, the skin underneath sure makes it look like a ladle, doesn't it?

This tells you that a pelican eats bigger things than a spoonbill, because a ladle can scoop up bigger things than a spoon. The student needn't include that last part.

Lesson 74

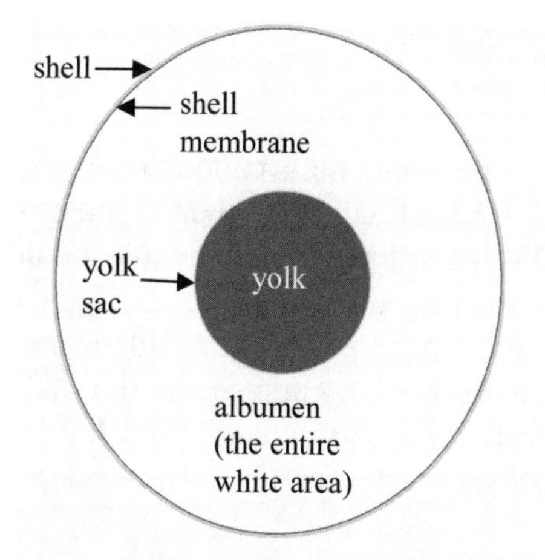

1. The yolk, yolk sac, albumen, shell membrane, shell
2. The bird embryo eats the yolk.

Older students: The drawing should look something like what you see on the right. The yolk is the food source for the embryo. The yolk sac holds the yolk. The albumen acts as a shock absorber, moistens the embryo, and is a partial food source. The shell membrane fights off bacteria, and the shell protects the egg contents.

Oldest students: The yolk will get smaller over time, because the embryo eats it.

Lesson 75

1. The curves help to spread out any pressure the egg gets, which help to prevent it from breaking, especially when its parent sits on it.
2. Incubation is the process of a parent bird keeping its eggs warm, usually by sitting on them.

Older students: The experiment showed that when you apply pressure evenly over an egg, it resists breaking. It usually breaks when pressure is applied only to one point on the egg. Since a parent sitting on the egg applies pressure evenly, it won't break the eggs it is sitting on. If a baby bird hatches blind and without feathers to keep it warm, it is altricial. Some birds are born with feathers and with the ability to see right away. They are precocial.

Oldest students: The kittens are altricial. They have a full coat of hair, but they cannot see. The horse is precocial. It can see right away and has a full coat of hair to keep it warm.

Lesson 76

1. It is an animal that people care for and use for food, clothing, work, or companionship.
2. They live on land.

Older students: Farm animals and pets belong on the "cattle" page, because "cattle" probably refers to all domesticated animals. The student should note that domesticated animals are animals that people care for and use for food, clothing, work, or companionship. Spiders, worms, and things that crawl close to the ground should be on the "creeping things" page, and the large, wild land animals should be on the "beasts of the earth" page.

Oldest students: Generally, anything with fur is a vertebrate, as all mammals are vertebrates. Also, any reptile (like a snake or lizard) is a vertebrate, as is any amphibian (like a frog or salamander). Insects, spiders, worms, centipedes, etc., are all invertebrates.

Lesson 77

For the picture on page 236: The top animal is the centipede, and the bottom one is the millipede. You can tell this because the millipede has lots more legs.

1. An insect has six legs.
2. A millipede is the animal that has the most legs.

Older students: The drawing of the insect should have six legs and two antennae. The spider should have eight legs and no antennae. Since the spider has more than six legs, it is not an insect.

Oldest students: The centipede drawing should look like the top picture on page 236. Centipedes usually have fewer legs than millipedes. However, they both have a lot more than six legs, so they are not insects.

Lesson 78

1. The front end is called the anterior end, and the back end is called the posterior end.
2. It uses them to hold on to the ground so it can move.

Older students: The drawing should look something like the top picture on page 237. The anterior end is the one nearest the clitellum, and the posterior end is farthest from the clitellum. The dorsal side is the top, and the ventral side is the underside. An earthworm moves by anchoring its anterior end and then scrunching up its posterior end. It then releases its anterior end, anchors its posterior end, and stretches out its anterior end. It does this over and over again to move.

Oldest students: Earthworm poop is called an earthworm's **castings**. It is also called **vermicast**.

Lesson 79

1. Mammals are covered with hair, reptiles are covered with scales, and amphibians are covered with smooth, moist skin.
2. The temperature of a cold-blooded animal's insides changes depending on the weather. The temperature of a warm-blooded animal's insides does not.

Older students: The most common amphibians are frogs and salamanders. They are cold-blooded and covered in smooth, moist skin, and they use it as one way in which they breathe. Reptiles include lizards, snakes, alligators, and turtles. They are cold-blooded and are covered in dry scales that prevent water loss. Mammals would be anything covered in fur. The hair insulates them, which they need, since they are warm-blooded.

Oldest students: The other two types of vertebrates are fish, which are cold-blooded, and birds, which are warm-blooded. Fish are covered in scales and live underwater, while birds are covered in feathers.

Lesson 80

1. Mammals are insulated by their skin, hair, and fat.
2. Fat is an important part of certain structures in living organisms, and it is also used to store excess food.

Older students: In the experiment, the student placed bare fingers and shortening-covered fingers in cold water and then in hot water. In both cases, his bare fingers felt the temperature of the water more than his shortening-covered fingers. This shows that shortening, which is fat, is a good insulator, since it protected the student's fingers from temperature extremes. The shortening did not dissolve in the water because, like oil molecules, fat molecules are not attracted to water molecules. Fat is also used by animals to store excess food, and it is an important part of some structures in organisms.

Oldest students: The polar bear would feel the cold of the artic more, because since he is not getting the food he needs, his body will start using his fat for its store of energy. That will reduce the amount of fat in his body, which will also reduce the amount of insulation.

Lesson 81

1. People have the most in common with mammals.
2. People have the most in common with the great apes.

Older students: There are more characteristics than these, but here are some examples. Your child doesn't have to have all of these:

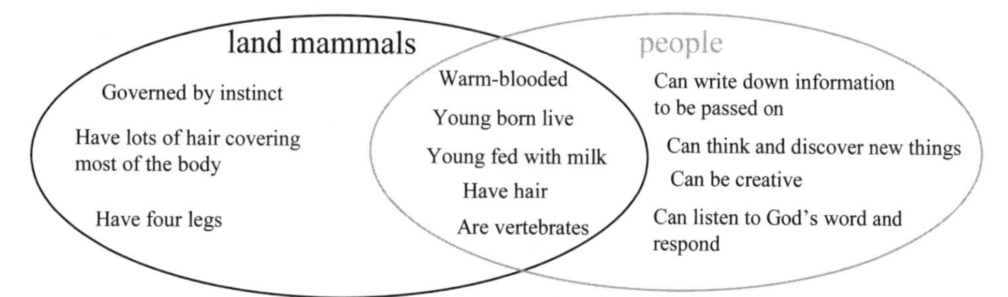

Oldest students: Whales are mammals that do not live on land. The student can indicate dolphins, but technically, dolphins are whales. Others would include manatees, dugongs, seals, sea lions, walruses, and otters. The student need not list them all.

Lesson 82

1. It can accurately determine how far things are from it.
2. No, it does not. A wide field of view sacrifices binocular vision, which is necessary for good depth perception.

Older students: The pictures should show one animal with eyes mostly on each side of the face, while the other animal has eyes mostly pointing forward. The forward-facing eyes allow for binocular vision. This lets the brain see the object with both eyes, which allows it to determine distance. That means there will be good depth perception. The side-facing eyes can't see many objects at the same time, which means the brain gets to see most things with only one eye, which results in poor depth perception. However, it provides for a wide field of view.

Oldest students: The drawing should have eyes that are farther apart and a bit more to the side than the animal with forward-facing eyes, but it should not have its eyes as far to the sides as the other animal. Pulling the eyes more to the side means there will be less binocular vision, which means worse depth perception. However, it will allow the animal to see more without turning its head, which means a wider field of view.

Lesson 83

1. If you can smell something, there must be chemicals in the air.
2. No, it does not. Insects smell with their antennae, and spiders smell with extensions on their legs.

Older students: The picture should look like the one on page 253, without the inner nose detail. Chemicals in the air must enter the nose, where chemical sensors can detect them and send information to the brain, which turns it into a smell. The nare is one of the big openings in the nose.

Oldest students: Some animals smell with their antennae or special structures on their legs. Other animals heighten their sense of smell by pulling in chemicals with their tongue and touching their tongue to a special organ that has even more chemical sensors.

Lesson 84

1. Sound is vibrations that travel through the air. The student could also have "a wave" in the blank.
2. The eardrum starts vibrating.

Older students: The picture should look like the one on page 256, minus all the detail. The auricle funnels the vibrations in the air (which is what sound is) to the auditory canal. When the vibrations reach the end, they make the eardrum vibrate. This makes the ossicles rock, which rocks the fluid in

the cochlea, which turns the fluid rocking into information that it sends along the cochlear nerve to the brain.
Oldest students: The one on the inside of the spaceship will hear the sound, because there is air in the spaceship. The astronaut outside will not hear the sound. Since sound is vibrations that travel through the air, if there is no air, there is no sound.

Lesson 85

1. They provide you with your static sense of balance, which tells you the position of your head.
2. They give you your dynamic sense of balance, which tells you how your head is moving.

Older students: Your static sense of balance works because there are tiny stones called otoliths in the vestibule of your inner ear. Those stones press against tiny hairs, and when your head tilts, they move. The hairs sense the stones moving, which sends information to your brain about the position of your head. Your dynamic sense of balance works because your semicircular canals are filled with fluid that moves as you move. The cupula bends in that fluid, which sends information to your brain about how the fluid is moving. This tells your brain how your head is moving.
Oldest students: The prefix "oto" refers to the ear. The rest of the word ("lith") refers to stones. Thus, an otolith is an ear stone.

Lesson 86

1. They are called taste buds.
2. The student can name any two of the following: sweet, salty, bitter, sour, and umami.

Older students: The taste buds are found in the bumps (or papillae) of the tongue. They can taste sweet, salty, sour, bitter, and umami. Every different combination of those five tastes results in a new flavor.
Oldest students: You will not lose your sense of taste forever. Your body continually makes new taste buds to replace the old ones. So even though you lost all your taste buds because of the burn, your body will eventually make new ones.

Lesson 87

1. It comes from your skin. The student could also say it comes from the receptors in your skin.
2. It exists all over your body rather than coming from a specific organ.

Older students: The student can get very creative with this. However, the basic idea is that the receptors are telling the brain what they are feeling. The hand in ice water, for example, has cold-temperature receptors that say, "This is really cold." The hot-temperature receptors don't say anything. Eventually, the cold-temperature receptors say, "I'm tired. Just assume I am still telling you it's really cold." Then they stop talking. When that hand gets put in the lukewarm water, the cold-temperature receptors are too tired to talk at first, but since the hot-temperature receptors are well-rested, so they immediately start saying, "It's hot." Eventually, however, the cold-temperature receptors wake up, and the brain finally gets receptors telling it that it is both hot and cold. That makes it realize the water is lukewarm.
Oldest students: The water will feel warmer than before. The cold-temperature receptors start going crazy when you get out of the water, but they eventually turn off. Thus, when you get back into the water, they are still turned off. As a result, your brain only hears from the hot-temperature receptors for a while, making the water feel warmer than it is.

Lesson 88

1. It is called the nasal cavity.
2. You can't smell well when you have a cold, so your sense of smell can't combine with your sense of taste to produce the flavors you are used to tasting.

Older students: The picture should look a lot like the one on page 268, but dots should exist on the tongue and going up into the nasal cavity, as if they were carried there by the breathing process. The student should discuss how particles from the food you are eating get moved into your nasal cavity by breathing, and that allows you to smell them. The smell then mixes with the taste coming from the tongue to produce the flavor you experience.

Oldest students: You can't smell well when you have a cold, so your sense of smell can't combine with your sense of taste to produce the flavors you are used to tasting.

Lesson 89

1. We call it the iris.
2. The opening of the iris is the pupil, and it changes size to allow different amounts of light into the eye.

Older students: The two pictures should look the same except for the size of the pupil. The pupil should be much larger in the dim light. The pupil changes size because the iris controls how much light gets into the eye. When there is little light, the iris opens up, making the pupil large. This lets in a lot of light. When there is a lot of light, the iris constricts, making the pupil small. This limits the amount of light that comes in.

Oldest students: An owl's iris opens wider because it needs to see well in very dark conditions. As a result, it needs to let a lot more light into its eyes.

Lesson 90

1. It is called an optical illusion.
2. We call it the dominant eye.

Older students: The picture should be a drawing of a palm with a hole in it and something showing through the hole. This was caused by the fact that the tube on the right eye and the palm in front of the left eye forced the eyes to be looking at two different things. This sent two different messages to the brain, which confused it. As a result, the brain constructed the optical illusion of a hole in the hand. Your dominant eye is the one that focuses your gaze. The student determined his or her dominant eye in the second experiment.

Oldest students: If the person started out with the same leg most of the times the student observed him or her, then that is the person's dominant leg. If the person used one leg half the time and the other leg the other half of the time, the person doesn't have a dominant leg.

TESTS

The First Day of Creation: Test #1 (Lessons 1-6)

1. What does it mean for light to reflect off something?

    a. Light hits something and breaks it.
    b. Light hits something and "bounces" off it, changing direction.
    c. Light hits something and passes through it.
    d. Light hits something and heats it up.

2. What must happen in order for us to be able to see an object?

    a. Light must be absorbed by the object.
    b. Light must pass through the object.
    c. Light must reflect off the object and hit our eyes.
    d. Light must be focused by the object.

3. Starting from the top of a rainbow and working your way down, list the seven colors of light in the order you find them.

4. If I were to take light of each color you listed above and mix them together, what color of light would I have?

5. An object absorbs all colors of light but green. What color is the object?

6. An object is red. What colors of light does it absorb? (use the colors of the rainbow to state your answer).

7. When an object absorbs radiant energy, what kind of energy is it changed into?

8. What kind of energy is found in food?

    a. Chemical energy
    b. Mechanical energy
    c. Radiant energy
    d. Thermal energy

9. Which color object would get warmer while lying out in the sun: a black object or a white object?

10. When you look at the inside of a toaster, you see the wires glowing, and you can feel the heat. What two kinds of energy are you observing?

11. How does a white piece of paper reflect light?

    a. It reflects the light so that the information in the light is preserved.
    b. It reflects all the light in the same direction.
    c. It adds radiant energy to the light.
    d. It scatters the light in many directions.

12. Does a white object reflect all of the light that hits it? Answer yes or no.

13. Write down the Law of Energy Conservation.

14. You have an electrical generator that converts mechanical energy into electrical energy. All you do is turn a crank, and electrical energy is produced. You hook up the generator to a light bulb and start turning the crank, and the light bulb lights up. After a while, you get tired, so you slow down the speed at which you are turning the crank. What happens to the brightness of the light bulb?

15. Someone tells you he has made a device that takes a certain amount of energy and turns it into twice as much energy. Why shouldn't you believe him?

The First Day of Creation: Test #2 (Lessons 7-12)

1. Fill in the blanks: A battery stores energy as _________ energy but converts it to ________ energy so it can be used in devices like toys and flashlights.

2. Two batteries are used in a flashlight. What energy conversions take place?

   a. Chemical energy is converted to radiant energy.
   b. Radiant energy is converted to chemical energy, which is then converted to electrical energy.
   c. Chemical energy is converted to electrical energy, which is then converted to radiant energy.
   d. Electrical energy is converted to radiant energy.

3. Name a kind of light that your eyes cannot see.

4. What kind of light does a television remote use to control the television?

5. In the diagram on the right, what is the structure pointed out by the arrow labeled "5?"

6. In the diagram on the right, what is the structure pointed out by the arrow labeled "6?"

7. In what part of the eye are the rods and cones found?

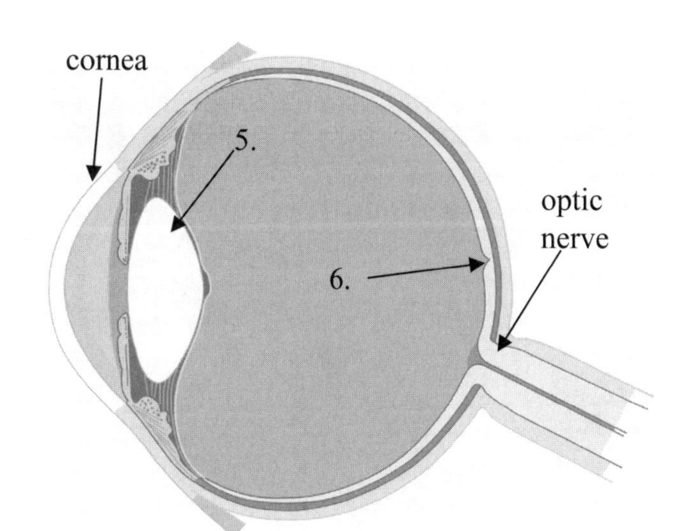

8. Explain why you sometimes see your reflection in a window.

9. Suppose you are looking outside through a window. You see a tree that is outside, but you also see your own reflection. Which of those things (the tree or your reflection) is the result of light passing through the glass?

10. When are you most likely to see your reflection in a window: when it is bright outside or when it is dark outside?

11. What is a one-way mirror?
   a. It is a mirror that reflects light only one way.
   b. It is a mirror that doesn't reflect light very well.
   c. It is a window that only allows light to pass through in one direction.
   d. It is a window that looks like a mirror on one side but acts like a window on the other side.

12. What makes a one-way mirror work?

   a. The reflective coating reflects light in only one direction.
   b. The lights are dim on the window side and bright on the mirror side.
   c. Light passes through it in only one direction.
   d. Light is reflected when it travels one way and transmitted when it travels the other way.

13. What do we call a cable that guides light so that it travels wherever the cable goes?

14. When light hits a transparent object, three things can happen to the light. What are they?

15. How can fiber optics be used by doctors?

The Second Day of Creation: Test #1 (Lessons 16-21)

1. What do we call it when a liquid turns into a gas?

2. What do we call it when a gas turns into a liquid?

3. How do clouds form?

    a. They form from tiny drops of water that rise from the ocean.
    b. They form from water freezing in the air.
    c. They form from water vapor that condenses onto tiny particles floating in the air.
    d. They form from water that evaporates from tiny particles floating in the air.

4. Does something need to be in a freezer in order to freeze?

5. Fill in the blank: To melt something, you must _______ thermal energy.

6. When water freezes, does it expand or contract?

7. When most things freeze, do they expand or contract?

8. What is volume a measure of?

    a. How heavy an object is
    b. How much room an object takes up
    c. How long an object is
    d. How tall an object is

9. Fill in the blank: In order to float, an object must weigh _______ than an equal volume of water.

10. Does ice float or sink in water?

11. Two cans of soda are put in water. One sinks, while the other floats. Which is the diet soda?

12. As you travel higher into the air, what happens to the temperature?

13. How can a drop of water in the ocean end up falling as rain on the roof of your house?

14. When water is attracted to the surface of a metal, is that adhesion or cohesion?

15. If water "beads up" on a surface, is it more attracted to the surface or to other water molecules?

The Second Day of Creation: Test #2 (Lessons 22-27)

1. If you took a droplet of water and continued to split it in half, eventually, you would reach something you could not divide in half. What would that be?

2. Why do chemists call water "H_2O?"

3. When you make a saltwater solution, what is the solvent?

4. For the question above, what is the solute?

5. You see a clear glass of water. There is nothing floating in it. Can you be sure it is pure water?

6. Name a gas that is dissolved in soda (like Coke).

7. Why does an aquarium need air pumped into it?

8. Does the phrase, "seeing is believing" work in science?

9. In one of your experiments, you pushed a cork down with a glass, but the glass never touched the cork. What was pushing the cork down?

10. When most things warm up, do they expand or contract?

11. You have a cold sample of air and a warm sample of air. In which sample are the molecules moving more slowly?

12. You see a blown-up balloon sitting in your freezer. You pull it out and notice it is very cold. If you leave it out for a long time, what will happen to the balloon?

   a. It will simply warm up to room temperature, and there will be no visible change.
   b. The balloon will warm up to room temperature, and it will get larger as it does so.
   c. The balloon will warm up to room temperature, and it will get smaller as it does so.
   d. The balloon will simply cool down, and there will be no visible change.

13. Does air have weight?

14. If a balloon floats in air, what can you conclude about the weight of the balloon and its contents?

15. In order to cause a hot-air balloon to sink slowly to the ground, what is typically done?

   a. The air is let out of the balloon.
   b. The contents in the balloon's basket are thrown overboard.
   c. The air inside the balloon is heated to a higher temperature.
   d. The heat is reduced so that the air inside cools to a lower temperature.

The Third Day of Creation: Test #1 (Lessons 31-36)

1. When a living thing decomposes:

   a. It stops all the processes that keep it alive.
   b. It takes in food and converts it to the energy it needs to stay alive.
   c. The molecules that make it up break down, turning into completely different molecules
   d. The molecules that make it up are reorganized to make the living thing stronger.

2. What is humus?

3. Which is the best description of soil?

   a. Soil is a mixture of humus and trash.
   b. Soil is a mixture of humus and many small bits of rock that come in many different sizes.
   c. Soil is a pure substance in which plants grow.
   d. Soil is composed only of decayed dead things.

4. (Is this statement True or False?) Plants grow better in soil that contains lots of humus.

5. If soil has large pores in it, it probably is made of a lot of:

   a. humus
   b. dirt
   c. small bits of rock
   d. large bits of rock

6. When rock is melted, and the liquid rock freezes, it forms:

   a. Igneous rock
   b. Metamorphic rock
   c. Sedimentary rock
   d. Soil

7. When small bits of rock come together and harden, they form:

   a. Igneous rock
   b. Metamorphic rock
   c. Sedimentary rock
   d. Soil

8. When sedimentary or igneous rock is put under a lot of heat and pressure, it forms:

   a. Igneous rock
   b. Metamorphic rock
   c. Sedimentary rock
   d. Soil

9. (Is the following statement True or False?) Freshwater is heavier than an equal volume of saltwater.

10. Suppose you put something in water and it sinks. What could you add to the water to make it float?

11. If you melt down an iceberg, will the water taste like the water in the ocean?

12. When you add a solute to a solvent, the temperature at which the solution freezes is lower than the temperature at which the solvent freezes. Scientists call this:

a. Freezing Temperature Decline
b. Melting Temperature Elevation
c. Protection Against Freezing
d. Freezing Point Depression

13. When ice has been in the freezer for a long time, is there any liquid water present?

14. (Is this statement True or False?) When salt dissolves in water, it slows down the rate at which water freezes.

15. (Is this statement True or False?) Adding salt to ice will melt the ice regardless of how cold it gets.

The Third Day of Creation: Test #2 (Lessons 37-42)

1. What is a hypothesis?

2. After you make a hypothesis, what should you do?

3. In a seed, what are the cotyledons for?

4. The book says a seed is a "baby plant inside a house with its lunch." What do we call the baby plant when it is in the seed?

5. Do all seeds have two cotyledons?

6. What part of the plant embryo develops into the plant's roots?

7. What is germination?

8. What must happen to the testa before germination can occur?

9. What part of the embryo develops into the plant's stem?

10. What is the function of a plant's stem?

11. Where is the epicotyl in relation to the cotyledons?

12. What does the plumule become?

13. What happens to the cotyledons once the plant has been growing for a while?

14. What three things does a plant need for photosynthesis?

    a. Water, soil, and light
    b. Soil, sunlight, and air
    c. Water, carbon dioxide, and light
    d. Carbon dioxide, soil, and light

15. What two things are made in photosynthesis?

    a. Water and air
    b. Glucose and oxygen
    c. Glucose and air
    d. Glucose and water

The Fourth Day of Creation: Test #1 (Lessons 46-51)

1. Why does a trees shadow change throughout the course of the day even though the tree doesn't move?

2. When is a tree's shadow the shortest?

   a. At noon
   b. In the evening
   c. In the morning
   d. At 2:30 PM

3. You mark the highest position of the sun in the sky in the middle of summer. You do it again in the middle of winter. Which time is the sun higher in the sky?

4. If it is made correctly, what can a sundial tell you?

5. A person measures his or her shadow at noon once each month. When the shadow is the longest, is it a winter month or a summer month?

6. Of the following colors, which is most likely to bounce off the particles in the air?

   a. Red
   b. Yellow
   c. Green
   d. Blue

7. (Is this statement True or False?) The sun's light must travel through more air in the evening than it does at noon.

8. (Is this statement True or False?) The sky is blue because it reflects the ocean, which is blue.

9. What causes it to look like the sun is moving across the sky?

10. Even when you sit perfectly still, you are actually still moving very quickly. Why?

11. How do we define the length of a day? (Don't tell me the number of hours. Tell me what determines that length of time.)

12. How do we define the length of a year? (Don't tell me the number of days. Tell me what determines that length of time.)

13. Which planet is closest to the sun?

14. Which planet is farthest from the sun?

15. Think about how long Jupiter takes to make one complete orbit around the sun. Does it take more or less time than the earth takes to make one full orbit around the sun?

The Fourth Day of Creation: Test #2 (Lessons 52-57)

1. Which are the outer planets?

   a. Jupiter, Mars, Uranus, Mercury
   b. Mercury, Venus, earth, Mars
   c. Jupiter, Saturn, Uranus, Neptune
   d. Uranus, Neptune, Mars, Mercury

2. Which is the largest planet in our solar system?

3. Which planets are rocky: the inner planets or the outer planets?

4. You go outside at night and see the entire surface of the moon lit up. What phase is that?

5. You go outside at night and see only half of the surface of the moon lit up. What phase is that?

6. What is the proper order of the following moon phases?

   a. Full moon, quarter moon, gibbous moon, crescent moon, new moon
   b. Full moon, gibbous moon, quarter moon, crescent moon, new moon
   c. Full moon, crescent moon, gibbous moon, quarter moon, new moon
   d. Full moon, new moon, crescent moon, gibbous moon, quarter moon

7. (Is this statement True or False?) The moon is larger when it is near the horizon because it is closer to the earth.

8. When the moon gets between the earth and the sun, what kind of eclipse results?

9. When the earth gets between the moon and the sun, what kind of eclipse results?

10. (Is this statement True or False?) A lunar eclipse only happens during a full moon.

11. (Is this statement True or False?) If star "A" burns more brightly than star "B," star "A" will always appear brighter than star "B" in the night sky.

12. (Is this statement True or False?) The sun is not the largest star in creation.

13. Why don't we see the stars during the day?

14. What is light pollution?

15. What is the most important difference between a planet and a star?

    a. A planet wanders in the sky, but stars stay fixed in their positions
    b. A planet is rocky, but stars are made of gas.
    c. A planet has life on it, but a star does not.
    d. A planet cannot make its own light, but a star does.

The Fifth Day of Creation: Test #1 (Lessons 61-66)

1. Which of the characteristics below are typically found in animals? (There is more than one answer.)

    a. Make their own food
    b. Can move from one place to another
    c. Must eat other living things
    d. Have roots

2. (Is this statement True or False?) The rules scientists make to explain God's creation always hold true.

3. (Is this statement True or False?) A freshwater fish is constantly absorbing water into its body.

4. You did an experiment where you put a potato in saltwater, and the potato got smaller. Why did it get smaller?

5. What word do scientists use to describe a barrier that lets some things cross through it but not other things?

    a. Semipermeable
    b. Partly transmissible
    c. Ruggedly partitioned
    d. Mostly blocking

6. Solutes tend to travel to places where there is a lot of solvent and not much solute. What do we call that process?

7. You put a drop of blue food coloring in a large glass of water and watch what happens over several days. Which of the following best describes what you see?

    a. The drop falls to the bottom of the glass and stays there. After several days, there is a glass of clear water with a blue drop at the bottom.
    b. The blue food coloring dissolves in the water, but it dissolves as a streak of blue. After several days, the glass will be mostly clear with a streak of blue where the drop fell in the glass.
    c. The blue food coloring dissolves near the bottom of the glass. After several days, the bottom of the glass will be dark blue, but the top of the glass will be clear.
    d. The blue food coloring dissolves in the glass. At first, parts of the glass will be dark blue and parts will be clear. After several days, the blue color spreads out, eventually making the entire glass the same shade of blue.

8. What do vertebrates have that invertebrates do not?

9. A clam has a hard, outer shell. Once you open that shell, everything inside is soft and squishy. Is the clam an invertebrate or a vertebrate?

10. Which are there more of in creation – invertebrates or vertebrates?

11. For what two purposes does a clam use its foot?

12. What is the name of the method an octopus uses to move quickly through the water?

    a. Water jetting
    b. Jet propulsion
    c. Jet streaming
    d. Stream jetting

13. What is the main job of the caudal fin on a fish?

    a. Steering
    b. Keeping the fish righside up in the water
    c. Propelling the fish through the water
    d. Slowing down

14. Some fish use the dorsal fin for protection, but all of them use it to:

    a. Steer
    b. Keep the fish righside up in the water
    c. Propel the fish through the water
    d. Slow down

15. When a fish wants to rise up to a shallow depth in the water, it:

    a. Puts oxygen in its swim bladder
    b. Takes oxygen out of its swim bladder
    c. Expels some of the contents of its stomach
    d. Eats more

The Fifth Day of Creation: Test #2 (Lessons 67-72)

1. (Is this statement True or False?) The shaft of a feather is solid so it is very heavy.

2. What part of a feather is like Velcro?

   a. The rachis
   b. The quill
   c. The shaft
   d. The barbules

3. What is the purpose of down feathers?

   a. They help the bird descend in flight.
   b. They help the bird ascend in flight.
   c. They provide insulation for the bird.
   d. They produce the shafts for the contour feathers.

4. What does the preen gland do?

   a. It makes oil.
   b. It make the feathers pretty.
   c. It digests food.
   d. It removes salt from the bird.

5. (Is this statement True or False?) A cormorant wants its feathers to get wet so it can dive deep underwater.

6. (Is this statement True or False?) When an airplane is flying, there is more air above the wing than there is below the wing.

7. Fill in the blank: The faster a wing is traveling through the air, the _________ the lift.

8. When a bird flaps its wings downwards, are they spread out or pulled in, close to the bird's body?

9. Fill in the blank: When a wing travels through the air, it generates lift. When a bird flaps its wings, it generates _______.

10. (Is this statement True or False?) When a large bird takes off, it usually faces into the wind.

11. Which of the following shapes is streamlined?

a. 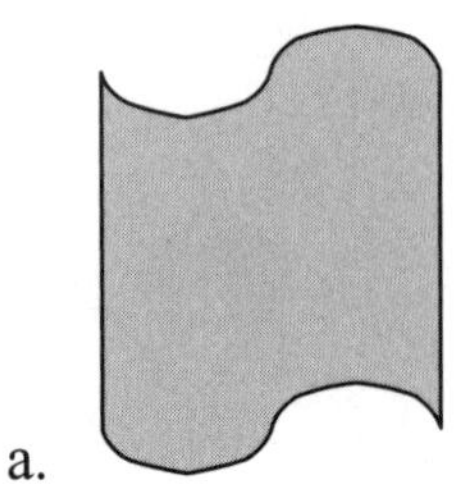

b. 

c.

d.

The owner of this book is free to photocopy this page

12. Why is it important for a bird's body to be streamlined?

13. Name at least two design features found in birds that allow them to fly.

14. (Is this statement True or False?) A bird's bones are mostly hollow.

15. You have two bones that are identical in size and shape. One comes from a bird, and the other comes from a cow. Which one weighs more?

The Sixth Day of Creation: Test #1 (Lessons 76-81)

1. What is a domesticated animal?

2. What are the three groups the Bible uses to classify land animals?

    a. Flying animals, swimming animals, crawling animals
    b. Cattle, creeping things, beasts of the earth
    c. Large animals, medium animals, small animals
    d. Vertebrates, invertebrates, nonvertebrates

3. How many legs do insects have?

4. (Is this statement True or False?) Spiders are insects.

5. Which has more legs: a centipede or a millipede?

6. What term do scientists use to refer to the front end of an animal?

7. What term do scientists use to refer to the top of an animal?

8. Fill in the blank: When an earthworm moves, it anchors one end of its body with its ________.

9. A reptile is:

    a. Cold-blooded and covered with fur
    b. Warm-blooded and covered with scales
    c. Cold-blooded and covered with scales
    d. Warm-blooded and covered with fur

10. There are at least three things that help insulate mammals. Name two of them.

11. Name a vertebrate that does not spend all its life on land.

12. (Is this statement True or False?) Plants and animals have fat in them.

13. What is fat used for besides insulation?

14. Do people have more in common with reptiles or mammals?

15. What makes people completely different from animals?

The Sixth Day of Creation: Test #2 (Lessons 82-87)

1. What does "binocular vision" mean?

    a. It means you can see things very far away
    b. It means you can magnify things with your eyes
    c. It means you can look at two different things at the same time
    d. It means you can see the same thing with both eyes at the same time

2. Fill in the blank: An animal with binocular vision will have good ________ perception.

3. Fill in the blanks: If an animal has eyes on the sides of its face, it has a wide _______ __ _______.

4. What does your nose detect in order to give you a sense of smell?

5. Does an animal need a nose in order to smell?

6. What is a Jacobsen's organ?

    a. It is an organ that enhances the sense of smell in some animals.
    b. It is an organ that enhances the sense of hearing in some animals.
    c. It is an organ that enhances the sense of sight in some animals.
    d. It is a musical instrument.

7. What vibrates when sound reaches the end of the auditory canal?

8. Can sound travel through space, where there is no air?

9. What are the stones in your ear called?

10. Which sense of balance (static or dynamic) do the stones in your ear give you?

11. Which sense of balance (static or dynamic) do the semicircular canals in your ear give you?

12. What are the five basic taste sensations?

    a. Rotten, Fresh, Sweet, Salty, Rancid
    b. Sweet, Salty, Bitter, Sour, Umami
    c. Fruity, Salty, Spicy, Fresh, Sour
    d. Savory, Fruity, Salty, Spicy, Fresh

13. What are the structures on your tongue that give you your sense of taste?

14. Where does your sense of touch come from?

15. (Is this statement True or False?) Your sense of touch is a special sense.

ANSWERS TO THE TESTS

The First Day of Creation: Test #1 (Lessons 1-6)

1. b
2. c
3. red, orange, yellow, green, blue, indigo, violet (You get this from Roy G. Biv)
4. You would have white light, because white light contains all the colors of the rainbow.
5. The object is green, as that is the only color it can reflect.
6. It absorbs orange, yellow, green, blue, indigo, violet.
7. It is changed into thermal energy.
8. a
9. A black object
10. You are observing radiant energy and thermal energy.
11. d.
12. no. Some of the light is absorbed.
13. Energy cannot be created or destroyed. It can only be converted from one form to another.
14. The brightness goes down. With less mechanical energy, there is less electrical energy and therefore less radiant energy.
15. You should not believe him because it goes against the Law of Energy Conservation, which says that you cannot create or destroy energy.

The First Day of Creation: Test #2 (Lessons 7-12)

1. chemical, electrical
2. c
3. The student can name infrared light, ultraviolet light, or microwaves.
4. Infrared light.
5. lens
6. retina
7. They are found in the retina.
8. Light can be reflected off the window, so the window can act like a mirror.
9. You see the tree because of light that passes through the glass.
10. You are most likely to see your reflection when it is dark outside so that the light passing through the window doesn't overwhelm your eyes.
11. d
12. b
13. It is a fiber optic cable.
14. The light can be reflected, it can be transmitted, or it can be absorbed.
15. They can be used to examine the inside of a person without opening the person up.

The Second Day of Creation: Test #1 (Lessons 16-21)

1. Evaporation
2. Condensation
3. c

4. No. Freezing just means a liquid turning into a solid. That can happen for some things (like wax) at room temperature..
5. add
6. expand
7. contract
8. b
9. less
10. It floats.
11. The diet soda is the one that floats.
12. It gets cooler
13. Clouds move as they collect water. If a cloud formed or moved over the ocean, it would collect ocean water that evaporates. It doesn't release its rain until it moves over your house, it would end up raining that ocean water on your house.
14. That is adhesion. Cohesion would be a water molecule being attracted to other water molecules.
15. It is more attracted to other water molecules than to the surface it is on.

The Second Day of Creation: Test #2 (Lessons 22-27)

1. It would be a molecule of water.
2. Because a water molecules is made of two hydrogen atoms and one oxygen atom.
3. Water
4. Salt
5. No, you cannot. If a solute dissolved in it, it could still look like pure water but be a solution instead.
6. Carbon dioxide is a gas dissolved in soda.
7. Fish breathe oxygen that is dissolved in the water. Pumping air into the water adds new oxygen to the water.
8. No, it does not. You can't see air, but it exists.
9. Air was pushing the cork down.
10. They expand.
11. They move more slowly in the cold sample of air.
12. b
13. Yes, it has weight.
14. The balloon and its contents weigh less than an equal volume of air.
15. d

The Third Day of Creation: Test #1 (Lessons 31-36)

1. c
2. Humus is the portion of soil that comes from decomposed dead things.
3. b
4. True
5. d
6. a
7. c
8. b
9. False
10. You can add anything that dissolves in water (to make a solution that weighs more) to make it float. Salt is the easiest thing to answer with, but anything that dissolves in water will work.

11. No. An iceberg is freshwater that has been frozen.
12. d
13. Yes, there is. There is always an equilibrium between water melting and water freezing.
14. True
15. False

The Third Day of Creation: Test #2 (Lessons 37-42)

1. It is an educated guess that attempts to explain something.
2. You should do an experiment (or gather data in some other way) in an attempt to confirm the hypothesis.
3. They provide food to the plant embryo.
4. It is called the embryo
5. No. Some plants produce seeds with only one.
6. The radicle develops into the plant's roots.
7. The process by which a seed grows into a plant
8. The testa must come off the seed.
9. The hypocotyl turns into the stem
10. The stem holds the leaves up and allows water to travel to the rest of the plant.
11. As the "epi" prefix tells you, it is above the cotyledons
12. The plumule becomes the plant's first true leaves.
13. They wither and fall off the plant.
14. c
15. b

The Fourth Day of Creation: Test #1 (Lessons 46-51)

1. The sun moves in the sky, changing the way the light hits the tree.
2. a
3. It is higher in summer
4. It can tell you the time of day.
5. It is a winter month. The sun is lowest in the sky in the winter, which means it casts the longest shadows then.
6. d
7. True
8. False
9. The earth's rotation makes it look like the sun is moving.
10. You are rotating with the earth. Since the earth rotates quickly, you move quickly as well.
11. A day is the time it takes for the earth to make one complete rotation.
12. A year is the time it takes for the earth to make one complete orbit around the sun.
13. Mercury
14. Neptune (Pluto is no longer considered a planet.)
15. Jupiter takes more time to orbit the sun, because it is farther from the sun.

The Fourth Day of Creation: Test #2 (Lessons 52-57)

1. c
2. Jupiter is the largest planet.
3. The inner planets are rocky.
4. That is a full moon.

5. That is a quarter moon.
6. b
7. False
8. That's a solar eclipse.
9. That's a lunar eclipse.
10. True
11. False – If the brighter star is much farther away, it will appear dimmer.
12. True
13. The sun's light is so bright that it overwhelms their brightness.
14. Light pollution is artificial light that brightens the sky enough to block out some of the stars.
15. d

The Fifth Day of Creation: Test #1 (Lessons 61-66)

1. b and c
2. False
3. True
4. It got smaller because the salt dissolved in the water pulled water out of the potato. The student could also answer, "osmosis."
5. a
6. It is called diffusion.
7. d
8. They have backbones.
9. A clam is an invertebrate.
10. There are many more invertebrates in creation than vertebrates.
11. It uses its foot to move and dig
12. b
13. c
14. b
15. a

The Fifth Day of Creation: Test #2 (Lessons 67-72)

1. False
2. d
3. c
4. a
5. True
6. False
7. greater (or any adjective that indicates the lift is larger)
8. They are spread out to increase the thrust.
9. thrust (The student could say "an upward push")
10. True
11. d
12. A streamlined body reduces air resistance, making it easier to fly.
13. Their feather shafts are hollow, their bones are hollow, they have wings, and they have streamlined bodies. (The student need list only two.)
14. True
15. The one belonging to the cow weighs more, because it is not mostly hollow.

The Sixth Day of Creation: Test #1 (Lessons 76-81)

1. It is an animal that people care for and use.
2. b
3. Insects have six legs.
4. False
5. Millipedes have more legs than centipedes.
6. The front end is the anterior end.
7. The top side is the dorsal side.
8. When an earthworm moves, it anchors one end of its body with its <u>setae</u>.
9. c
10. Skin, hair, and fat. (The student needs to name only two.)
11. Fish and birds are vertebrates that do not spend all their time on land. Amphibians will work as well. (The student need name only one, and the student can name a specific fish, bird, or amphibian, such as a carp, hawk, or frog.)
12. True
13. It is important in the structures of many organisms, and it is used to store excess food. (The student need mention only one.)
14. People have much more in common with mammals.
15. People are made in the image of God. (The student can also list characteristics that indicate the image of God, such as the ability to respond to the Word of God.)

The Sixth Day of Creation: Test #2 (Lessons 82-87)

1. d
2. An animal with binocular vision will have good <u>depth</u> perception.
3. If an animal has eyes on the sides of its face, it has a wide <u>field of view</u>.
4. It detects chemicals in the air.
5. No.
6. a
7. The eardrum vibrates.
8. No.
9. They are called otoliths.
10. They give you your static sense of balance.
11. They give you your dynamic sense of balance.
12. b
13. taste buds
14. It comes from your skin (or the receptors in your skin).
15. False